SPACE APPRENTICE ID

YOUR
PHOTO
HERE

NAME:

principia mission

The Principia Space Diary was first published in 2015 with support from the UK Space Agency and European Space Agency astronaut Tim Peake. This book is accompanied by teaching resources that use real-life science and engineering to inspire and engage children in STEM learning. For free videos and downloads, visit discoverydiaries.org/diary/principia-space-diary

This third edition published 2020
by Curved House Kids Ltd
60 Farringdon Road
London EC1R 3GA
www.curvedhousekids.com
info@curvedhousekids.com

Written by Lucy Hawking with Kristen Harrison
Academic input from Professor Peter McOwan
Illustrated by Ben Hawkes
Designed by Angelique Hering, Kristi Korotash and Alice Connew
Editorial Management by Lucie Stevens

A CIP record for this book is available from the British Library.

ISBN 978-1-913269-25-8

www.discoverydiaries.org

principia mission
SPACE DIARY

By Lucy Hawking,
the Space Crew and YOU!

Illustrated by Ben Hawkes

THE CURVED HOUSE kids

WELCOME, SPACE APPRENTICE!

On 15 December 2015 British ESA astronaut Tim Peake was launched into space on his historic space mission, called Principia. Tim had to work super hard before and during his mission – he had to get fit and healthy, prepare for life on the International Space Station, do experiments, study the Earth from space and so much more.

Now that Tim is back safely on Earth, he needs your help. He needs Space Apprentices like you to report on his adventure. You will be retracing his journey – from training to returning to Earth – and documenting his incredible mission. This is your Principia Mission Space Diary to log your findings. I hope this will be the start of your own adventures in space!

GOOD LUCK!
Lucy Hawking
and the Space Crew

SPACE GLOSSARY

Keep your eyes peeled for these words as you work through your Space Diary. What do they mean? Can you create a longer glossary with your own new space words?

Astronaut A person who travels into space! The Russian word for astronaut is 'cosmonaut'.

Gravity

Soyuz

Sokol

ISS

Principia

PRE-LAUNCH: ASTRONAUTS WANTED!

Astronauts have to be fit, healthy and fully prepared before they go into space. Tim was training for his mission for four years before he blasted off to the International Space Station (ISS)!

Complete this chapter to make sure you are fit to fly and ready to begin your Principia Space Apprenticeship...

ASTRONAUT WORKOUT!

This energetic workout will get you ready for space! Complete each exercise and record your results.

Can you think of other exercises that might help you prepare for space? Create your own workout and try it with your friends!

1. JUMP FOR THE MOON

How many jumps can you do in 30 seconds?

Result: 60

2. FLOAT

You need to get used to floating in space. Lie on your tummy and stretch out like an aeroplane. Can you hold this for 30 seconds?

 Yes ☐ No

3. BALANCE

Astronauts need good balance. How long can you balance on one leg? If you find this easy, try closing your eyes and blocking your ears!

Left leg: 13 (secs)

Right leg: 15 (secs)

4. STRETCH

Your body will grow in space! How high can you reach with your hands right above your head?

Result: _____ cm

5. BREATHE

You need to have a calm mind to make sure you can deal with any situation in space. Breathe in and out slowly for one minute. Are you relaxed and ready to fly?

 Yes, let's fly! No, try again!

YOUR BODY IN SPACE

Zap for answers!

Hey guys!

I'm Marco and I help astronauts like you understand how your body will change in space. Can you help me work out which five of these are true?

10. Add your own question and test your friends!

WHEN YOU GO TO SPACE YOU MAY FIND THAT...

	True	False
1. You get taller.	☒	☐
2. You'll probably feel sick for your first few days in space.	☒	☐
3. Your thumbs fall off.	☐	☒
4. Your eyeballs change shape.	☒	☐
5. Your bones become weaker.	☒	☐
6. Your ears turn purple.	☐	☒
7. You grow hair all over your body.	☐	☐
8. Your face gets puffy.	☐	☐
9. You can't poo in space!	☐	☐
10.	☐	☐

SPACE DINNER

MENU

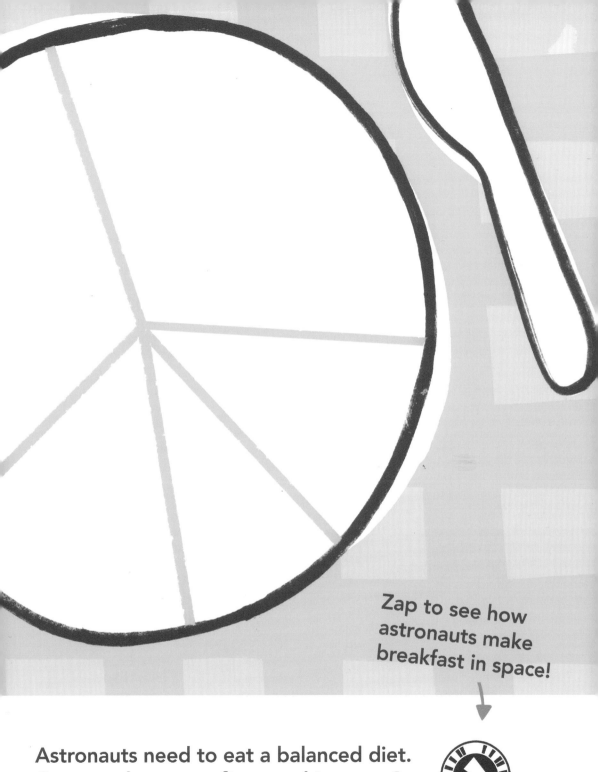

Zap to see how astronauts make breakfast in space!

Astronauts need to eat a balanced diet. Can you draw your first meal in space? Make sure it has all the right nutrients for a healthy space traveller!

DESIGN YOUR SPACESUIT

It's time to design your own spacesuit!

You can design it however you want but don't forget to include features that will let you breathe, communicate and keep you protected.

Zap to see Tim in
the Sokol suit he
wore for launch
and re-entry and
find out what
special features
your suit will need.

ASTRONAUT QUIZ!

What have you learnt about being an astronaut? Create your own quiz and test your friends!

True
OR
false?

CHAPTER ONE: GOODBYE TO EARTH!

Welcome to the launch pad!
It's time to leave the
Earth behind.
Get ready for liftoff –
we are going to retrace
Tim Peake's journey to space!

The countdown begins now...

TIME FOR LAUNCH!

Vinita here! It's 15 December 2015, Tim Peake's launch day. Help us tell his story. Can you add the times to the clocks and draw the missing scenes?

Zap for answers!

It's Principia Launch Day! Tim has arrived at the launch site right on time:

8:33 AM

20 minutes later... Tim enters the Soyuz

"Mission Control to all vehicles: It's 10:48am, please evacuate the launch site and get ready for launch!"

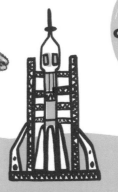

11:03 AM

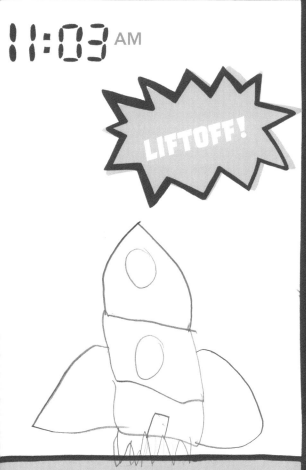

LIFTOFF!

9 minutes after launch the Soyuz separates from the rocket

Wow! 6 hours and 10 minutes later the Soyuz meets the ISS

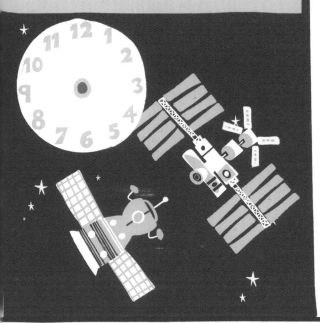

 PM

At 7pm the hatch opens and Tim boards the ISS just in time for dinner!

Zap and listen to U.S. astronaut Richard Garriott describe how it feels to launch into space

8 MINUTES TO SPACE

Liftoff! It took Tim Peake and his crewmates, Tim Kopra and Yuri Malenchenko, about 8 minutes from launch to orbit. Imagine you are there too. How does it feel blasting off into the blackness of space for the first time?

As our spacecraft lifts off I feel...

FAST-TRACK RENDEZVOUS

The Soyuz  and the ISS must time their meeting in space perfectly. This is called a "rendezvous". The Soyuz carrying Tim Peake and his crewmates did a fast-track rendezvous. Can you help us write up a report of the details?

Trace the route the Soyuz took to find out how many times Tim orbited the Earth.

(Use the **TIME FOR LAUNCH** cartoon at the start of this chapter to help you find some of the answers.)

SOYUZ TMA-19M
RENDEZVOUS REPORT

Zap for answers!

Tim Peake and his crewmates, _TIM Kopra_

from the USA and _Yuri Malenchenko_

from Russia, were launched at exactly __ __ : __ __ on

15 _____ 2015. Their spacecraft orbited

the Earth _____ times and it took exactly ____ hrs

____ mins from launch to docking. This is called a

fast-track _____.

CHAPTER ONE
WORD SEARCH

Zap for the answers!

Find the words you've learnt in this chapter.
Words can go forward, backward and diagonally.

U	N	L	W	Z	T	N	N	T	R
W	J	E	L	K	O	T	M	F	B
U	B	T	J	I	F	I	N	A	H
T	J	Z	S	M	L	B	A	R	Q
W	E	S	V	A	J	R	X	C	M
Z	I	K	U	Y	A	O	S	E	S
M	S	N	C	Y	J	V	J	C	O
G	C	I	R	O	C	E	A	A	Q
H	R	O	V	H	R	V	F	P	Y
D	S	O	Y	U	Z	U	Z	S	X
U	P	G	V	S	T	I	J	H	U

**Can you find 6 words
beginning with these letters?**

L M O R S S

CHAPTER TWO: SPACE CHAT

Which languages do astronauts speak
on the ISS? How can you tell the world
what it's like to live in space? And what
happens when a mysterious message
arrives, written in code?

UNITED IN SPACE

Welcome to the International Space Station!
Which countries have sent astronauts to the ISS?
Can you make a country card for one of them?

EUROPEAN SPACE AGENCY (ESA)

Country/Region: Europe

Greeting/s: Hallo (German), Hello (English)

First astronaut: Ulf Merbold (Germany), 28 November 1983

Time in space: Samantha Cristoforetti (Italy) had one of the longest single flights by a woman: 199 days, 16 hours.

Awesome fact: Tim Peake was the first British astronaut to fly with ESA.

Country/Region:

Greeting/s:

First astronaut:

Time in space:

Awesome fact:

Zap to meet some of the astronauts who have been to the ISS.

Hallo! Privyet! Konnichiwa! It's time to meet your fellow astronauts from all over the world.

BREAKING NEWS!

Not everyone gets to go to space, so it's important that you share your experiences with people on Earth.

I'm **Cindy** and I help people learn about space and science here on Earth. Can you write a news report about Tim's first day in space? Include a picture too!

DAILY ISS

DATE :

EARTH TO PRINCIPIA

Hi, I'm Berti, Mission Director at the European Space Agency. It's my job to keep the contact between Earth and the ISS. I've just received this message! Can you help me decode it?

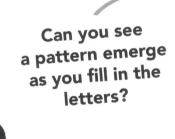

Can you see a pattern emerge as you fill in the letters?

Zap for the answer!

A	D
B	E
C	F
D	G
E	H
F	I
G	J
H	K
I	L
J	M
K	N
L	O
M	P
N	Q
O	R
P	S
Q	T
R	U
S	V
T	W
U	X
V	Y
W	Z
X	A
Y	B
Z	C

Ebiil!
Fp / qefp / Mixkbq / Bxoqe?
Fq'p / x / ybxrqfcri /
sfbt / colj / rm / ebob.
Qfj / Mbxhb,
pfdkfkd / lcc!

He_l_l_o_!
_I_s/t_H_is / P_l_an_et_ / _e_A_r_t_h_?
_I_t_'_s_ / _A_ / b_ea_U_t_i_fu_l_ /
v_i_e_w / f_R_o_M_ / _u_P_ / _H_e_r_e_.
_T_i_m / _p_e_a_k_e_,
_S_i_g_n_i_n_g/ o_f_f_!

CHAPTER TWO
WORD SEARCH

Zap for the answers!

Find the words you've learnt in this chapter.
Words can go forward, backward and diagonally.

F	M	O	E	H	W	M	R	A	A
Z	J	A	T	W	E	K	S	I	B
G	Q	R	G	S	R	T	P	T	M
B	A	V	S	R	R	I	C	N	V
E	Z	A	D	O	C	A	R	D	V
F	G	E	N	N	T	G	J	J	O
E	U	A	I	N	H	B	F	V	P
H	U	R	O	E	D	O	C	E	D
T	P	C	L	R	J	Z	J	B	F
U	G	S	J	E	Q	B	G	I	E
U	B	V	C	E	W	X	R	A	Q

**Can you find 6 words
beginning with these letters?**

A C D E M P

CHAPTER THREE: I SPY...

Your new home, the International Space Station (ISS), is a complicated structure that has been carefully built in space itself. Get to know the ISS by exploring all the parts of the spacecraft and creating your own diagram. And don't forget the windows, there's lots to see from space!

YOUR NEW HOME

Your new home is the International Space Station. This amazing structure was built in space using components flown up from Earth. If you look closely, you can see lots of shapes in the ISS that you can find on Earth too.

Each part of the ISS has its own special role.
Can you work out what each one does?

The "trusses" are the long beams and triangles
that hold the ISS together. The triangles provide
Support and make sure that the trusses can
hold the huge ISS together.

The sphere and canister shapes are the areas where
the astronauts _____ and _____
These are pressurised, just like a can of fizzy drink!

The large rectangular panels are solar panels that
collect _____ and turn it into _____ .

The smaller rectangular panels are thermal radiators
that get rid of _____ that the ISS produces.

The Robotic Arm is long and bendy so it can reach
around the ISS and fix things on the outside. It can
move nearly 100,000 kilograms of equipment! If an
elephant weighs 5000 kg, how many elephants could
the robotic arm lift? _____.

**Zap for
the answers!**

DRAW YOUR OWN ISS

Now that you know the ISS inside and out, we need your design and engineering skills to create a diagram of it.

Follow the numbers to make the outline, then colour all the components with your own colour key.

KEY

 Trusses for support

 Living and working modules

 Solar Panels for producing energy

 Thermal radioators to get rid of heat

 Robotic Arm for maintenance

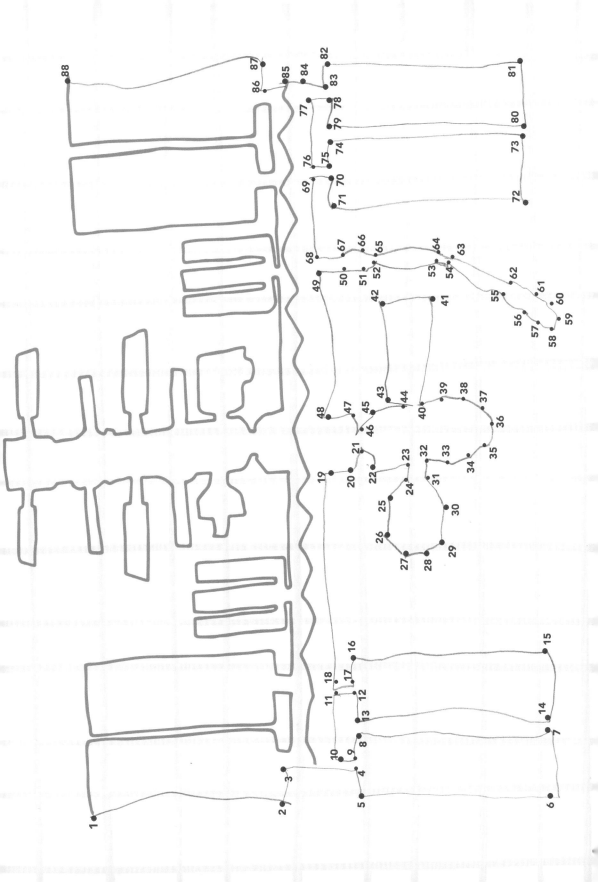

LOOKING AT THE EARTH FROM SPACE

Tim had an amazing view of Earth from space. He could see oceans, mountains, deserts and cities. Now it's your turn to observe the wonder and beauty of our home planet!

Zap the code below to see some of Tim's **incredible** photos of Earth from space. Choose one to explore and write a travel blog about your discoveries.

MY TRAVEL BLOG

THE SOLAR SYSTEM

Colour the planets in the Solar System, then write a planetary report describing the conditions on each.

This will help you learn which planets are explorable.

MERCURY

EARTH

VENUS

MARS

CHAPTER THREE
WORD SEARCH

Zap for the answers!

Find the words you've learnt in this chapter.
Words can go forward, backward and diagonally.

T	R	U	S	S	E	S	T	V	X
E	R	U	T	C	U	R	T	S	F
S	K	L	I	D	S	N	C	J	D
T	F	E	Q	O	M	I	W	T	D
E	Y	D	Y	F	T	I	X	V	Z
N	V	U	P	O	V	X	B	T	W
A	S	R	B	N	E	H	P	Q	Q
L	T	O	E	U	G	U	T	S	D
P	R	B	E	S	Y	E	E	O	Y
D	Q	E	I	K	B	O	L	M	H
V	R	A	J	H	C	O	V	A	D

**Can you find 6 words
beginning with these letters?**

K O P R S T

CHAPTER FOUR: SPACE FOR SCIENCE

The ISS is not just a spacecraft and home to astronauts, it is also a laboratory where astronauts do space experiments. Can you help Tim with some science and discovery?

SPACE GARDENING

During the Principia mission, students across the UK helped Tim by growing rocket from seeds that Tim had taken to space. Did you know that scientists from the European Space Agency have identified nine other foods that are suitable to grow in space? Zap here to find out what they are!

I'm Libby, and I'm in charge of coordinating the experiments that happen on the ISS. Can you find out which foods are suitable to grow in space? Then you can create your own space laboratory and try growing one of these plants in real life! Draw and label your plant here.

MAKE A SPLASH IN SPACE!

Astronauts need water just like we do on Earth – to clean their teeth, wash their hands, drink and prepare food. Zap here to see how Tim used water on the ISS.

Draw or write the lifecycle of a drop of water to show how astronauts use and reuse this precious liquid on the ISS. Start with one of the astronauts taking a drink. What happens next?

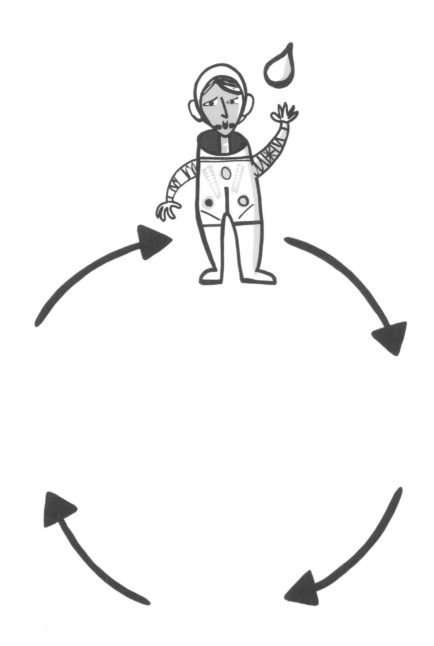

EXPERIMENTALLY YOURS

Tim did lots of experiments on the ISS, like spinning around really fast to see if he would feel dizzy in space (zap to watch the video!). Now it's time for you to do your first experiment.

I want to find out...

_ _ _ _ _ _ _ _ _ _ _ _ _ _ _

I will need the following materials:

_ _ _ _ _ _ _ _ _ _ _ _ _ _ _

My method will be to:

_ _ _ _ _ _ _ _ _ _ _ _ _ _ _

I predict...

_ _ _ _ _ _ _ _ _ _ _ _ _ _ _

Draw and label a
diagram of what
your experiment
would look like.

CHAPTER FOUR WORD SEARCH

Zap for the answers!

Find the words you've learnt in this chapter.
Words can go forward, backward and diagonally.

Y	Z	E	U	A	Y	M	N	E	Y
U	R	Q	L	Q	A	A	D	E	R
M	B	O	G	Z	N	M	L	S	E
V	E	J	T	Z	G	C	G	K	V
Y	G	T	W	A	Y	S	W	J	O
B	Y	D	H	C	R	W	Y	K	C
O	Q	Y	E	O	V	O	I	I	S
B	V	F	G	U	D	A	B	I	I
D	I	A	G	R	A	M	H	A	D
L	P	A	E	D	A	J	N	D	L
T	N	E	M	I	R	E	P	X	E

**Can you find 6 words
beginning with these letters?**

D D E L L M

CHAPTER FIVE: TO BOLDLY GO

You've made it to the ISS,
observed the Earth from space,
run experiments with Tim, so now...
where are you going next?

MAKING HISTORY

Nothing

1950   1960   1970   1980

1961:
First human
spaceflight!

_____ :
Voyager probe
leaves Earth
for outer Solar
System

1969:
Astronauts
land on the
Moon
(moonah)

1957: 'Sputnik',
the first
Spaceman
to orbit Earth

1986:
The Russian/Soviet
Mir Space Station
is launched

The first astronauts arrived on the ISS in 2000. How many years before you were born was that? Finish the timeline, then add the birth dates of your family and friends (and yourself) in the top half to find out where you fit in space history.

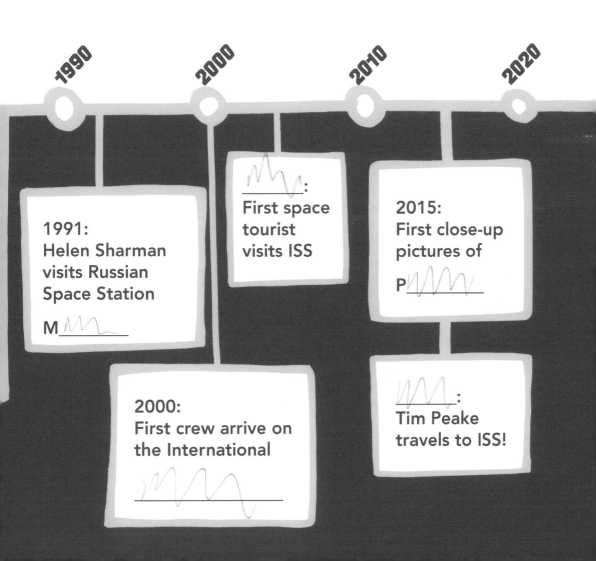

1990 2000 2010 2020

1991:
Helen Sharman
visits Russian
Space Station

M_____

_____:
First space
tourist
visits ISS

2000:
First crew arrive on
the International

_____

2015:
First close-up
pictures of

P_____

_____:
Tim Peake
travels to ISS!

SPACE HABITAT

It's time to explore your cosmic neighbourhood. Go back to your Solar System report and choose a planet to make your home. Draw your space city here and don't forget to include things that will help you survive the conditions on your planet.

ROBOTS IN SPACE

Robots are really useful in space. They can do lots of different jobs, some of which are too difficult or dangerous for humans. Which kind of rover will you use to explore your new home?

SPACE PROBE
Discovery Device

MARS ROVER
The Brave
Explorer

ROBONAUT
Your Robotic Friend

SPACE TELESCOPE
The Long Distance
Photographer

MOON BUGGY
The People Mover

MASS TRANSIT LUNAR VEHICLE
The Space Bus

CHAPTER FIVE
WORD SEARCH

Zap for the answers!

Find the words you've learnt in this chapter.
Words can go forward, backward and diagonally.

P	B	G	R	E	P	O	R	T	C
P	T	M	V	N	C	L	I	M	X
G	F	H	I	I	S	Q	M	K	Y
E	W	R	M	L	W	E	H	L	K
Q	C	S	I	E	D	A	P	P	J
K	O	I	W	M	B	C	R	U	S
C	O	N	D	I	T	I	O	N	S
F	Z	U	T	T	M	A	B	L	U
J	Q	A	U	T	X	A	E	U	K
I	T	C	Y	A	L	W	N	W	N
C	P	T	V	R	Q	H	F	N	L

**Can you find 6 words
beginning with these letters?**

C C H P R T

CHAPTER SIX:
MISSION FINALE

Astronauts! It's time to go.

**Your mission is coming to a close,
but you still need to land safely on Planet Earth.
Follow the next steps to plot your route home and
prepare to become an Earthling once more...**

RE-ENTRY

It's time to leave the space station! You've undocked the Soyuz capsule from the ISS and you're getting ready for descent. Hold tight – it's going to be a bumpy ride.

I'm Richard, and I'm a space junk expert. Help me guide the re-entry capsule back to Earth without hitting any floating debris.

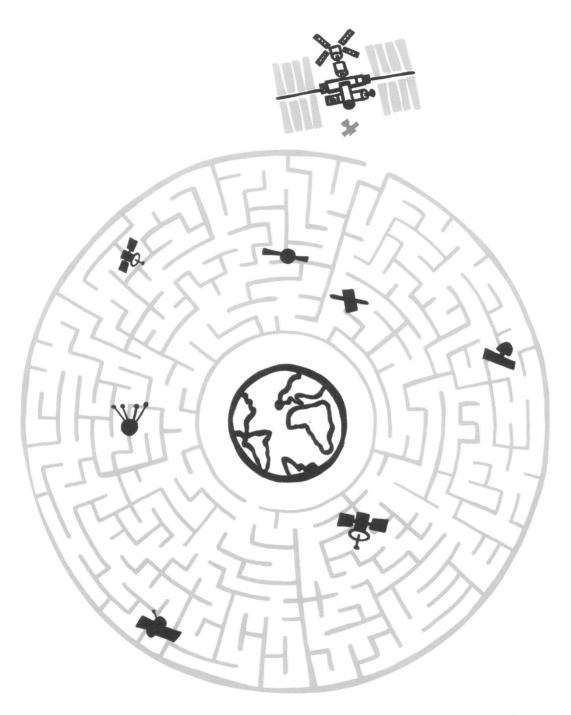

Zap for the answer!

THE JOURNEY HOME

On 18 June 2016, Tim's Soyuz capsule landed safely in a desert in Kazakhstan. Tim was transported to the Cosmodrome in Baikonur, then to Cologne, Germany, where he spent his first night back on Earth. Where will you land? On solid ground or out at sea? Draw a map of the route from your landing site to home.

Zap to watch
Tim's landing!

selffaw

N

eve

W

E at

S

soggy

SEND A POSTCARD TO SPACE

What a mission! You're back on Earth and you'll be bursting to tell your family and friends about your mission to space. But first, write a postcard to your fellow astronauts on the ISS to let them know you're back.